Dear Daddy:

True Child Abuse Stories that will make you cry

Short stories on Child Abuse based on true events

1. Infant daughter Masha abused physically and carried disturbing injuries due to sadistic nature of father
2. Physical abuse, Torture and first degree burns suffered by 10 month old baby girl Lynda
3. Inhumane physical abuse leading to disabilities suffered by 5 month old baby John who nearly escaped death
4. Dana, 3 year old toddler locked and raped by father repeatedly for 4 years
5. Baby girl Riah brutally assaulted and mistreated by her own father causing long term trauma
6. Emotional and physical abuse suffered by 5 year old Jerry for lack of concentration in studies
7. Heartless Dad tried to sell his own 3 month old daughter
8. 9 month old baby girl tormented physically and carried injuries which left her paraplegic
9. Daughter sexually abused by father repeatedly for 2 years
10. 6 year old son physically abused by father every day for 2 years leaving young Preston traumatized forever.

Infant daughter Masha abused physically and carried disturbing injuries due to sadistic nature of father

On 5th august, 2011 Masha was born in Sacred Hospital. Her mother Jane felt happy that their baby was born safe and sound. Soon after her birth they had to move from their hometown to California as her husband lost his job and jane got a new job from her friend's reference. Masha's father Wayne became full time caretaker for his 1 month old infant daughter as jane was the sole bread winner.

Wayne was disturbed mentally for being jobless and he started drinking. Whenever jane was not home he used to show all his anger on his little daughter. Neighbors heard her cry all day long as she was not fed and cleaned properly. Her father would squeeze her with so much force till her face would become red and eyes turned reddish due to rush of blood. He had an urge to always beat and smash her face. When Jane noticed that Masha was having scares all over her body Wayne confessed that he was not able to handle her and he apologized to her. Jane thought that he would change by the time Masha becomes older. But she was not aware that his urge to beat her infant was increasing day by day.

 Whenever Masha cried her father would shake her up with so much force that she slept unconsciously. She was 2 year old when she was brutally kicked, beaten up and choked by her father. He even pushed his finger all the way down to the child's throat till her mother came home and rushed her daughter to the hospital. Wayne was charged with criminal activity and attempt to murder. All his family members and neighborhood were disturbed by his pitiless activities towards his own daughter.

When he was investigated he claimed that his urge to smash and break his daughter intensified for last 1 year and whenever he woke up with a hangover to change his infant's diaper she was squeezed in the thighs till he heard clicking of her bones. Doctors found that 2 year old Masha was having difficulty in breathing, ribs were damaged, right hand was fractured and severe injury in the spinal cord. They also found bruising in child's private parts and Wayne confessed that he was tired of changing his child's diaper, so he used to harm her in anger. Even her mother was charged with act of negligence and jailed for 2 years as she didn't try to save her daughter from physical abuse.

Now Masha is with her foster parents living in a healthy environment and being loved unconditionally which every child deserves.

Introduction

Each and every day Children are being abused in one way or the other by family members or strangers. One cannot dismiss the fact that a child needs both mother and fathers love to live in a healthy environment. A father is a person who guides, protects and nurtures his child. What if he himself is responsible for abuse and violence against his child? The person who is responsible for his child's safety performs inhuman actions against his own kin then how can anyone protect their children from mistreatment by strangers. Among the Children who are abused around the world half of the child abuse cases come from their father.

Childhood enjoyment and innocence creates a building block for personality development and growth. It's a shocking fact that even newborn babies suffer physical abuse and lifelong injuries. In worst cases children have even died due to their father's negligence, physical violence and sexual abuse. Children who are abused by their fathers feel unprotected and insecure. It's hard for them to trust anybody. Abuse by their parents traumatizes them deeply in many ways. Neglected children suffer from betrayal, mistreatment, fear and mistrust. They feel alone, anxious and not worthy of living.

Based on surveys, it was found that most of the child abuse case by parents is happening between the age group of 1 month to 8 years. And 75% of the cases are from a child's father. Compared to boys girls are frequently abused from the age of 3 years. The child who experiences Physical abuse and violence goes through a transformative phase. When a child is abused in their development stage it affects their mental state. Their trust and love is shattered when they are mistreated by their own parents.

Most of the times children are threatened by their parents and they feel ignored as they are not able to talk to anybody about their problems. What about newborn babies who suffer physical abuse by their parents? They are unable to fight back, scream or move away when they are being beaten brutally by their father. It's a shameful fact that most of the abusive cases occur with these small babies who are scarred for their lifetime.

Affected children may show signs of abuse through their behavior, school performance, sleeping patterns, reduced development and coping ability with new people or situations. If we pay a little attention to all these symptoms we could prevent child abuse cases and save the innocent children from lifetime of suffering, fear and betrayal. The children who are affected never open up and talk about their abusive parents easily. People around them should engage with them frequently and support them.

Children should feel that they are always loved by people around them. Every child is precious and we could stop child abuse by identifying some signs of depression, attitude and fear shown by the children. These child abuse cases by father shows us how children are being mistreated around the world and what kind of consequences these children face due to the negligence of their parents and family members.

Physical abuse, Torture and severe burns suffered by 10 month old baby girl Lynda

Lynda, 10 month old baby girl was the only daughter of Mark and Maggie cooper. As Maggie wanted to continue her work they decided that mark would become the main caretaker for their daughter because he was working as a part time electrician.

One day Mark called his wife Maggie asking her to hurry home as their daughter slipped in hot water and burned herself while he was giving her bath. Soon child welfare officials intervened and took custody of baby Lynda who was found to be having "Pseudo abusive" and Scald burns as reported by the doctors.

Doctors were taken back by surprise and agony as they had not seen such a case in their life on severe abuse of babies. Lynda's both legs were covered with scalding burns and injured so painfully that she couldn't move her legs. Her whole neck and body were full of bite marks and burns inside her mouth. She was also suffering from brain injury due to shaken body syndrome. As if her suffering was not enough mark placed her on charcoal grill just because she was crying a lot. Even investigators were shocked by the heights of his abusive nature towards a small baby who doesn't even know why she is being punished by her own father who is supposed to love and care for her.

Police found out that Lynda was shaken regularly and injured severely. When mother Maggie wanted to take her to hospital mark talked her out of it by threatening her that child care officials would take their daughter away from her. Two days later Maggie noticed that her baby girl was carrying dark circles around her neck, bruises all around her body and scratches on her legs. Maggie was not able to look at her daughter going through such pain, she immediately admitted Lynda in children's hospital.

According to the medical report it was found that baby Lynda was severely burned over her lips, legs, had bite marks and brain injuries. 15% of her body was covered by burns. Investigators found that her father used to slam her down and violently shake his daughter regularly to make her sleep. The burn on her back was found to be from the charcoal grill in their kitchen.

Lynda was so traumatized that she was not able to sit up and stopped making eye contact with anyone. She gradually became lethargic. She was not even one year old and she had to go through torture and trauma from her own father. This kind of inhuman activity and torture shown to this child for several months was not taken lightly by the judge. Lynda's mother Maggie, 23 was sentenced to 3 years of prison for not preventing physical abuse and child neglect. Her father mark cooper was sentenced to only 30 years of imprisonment for scarring his baby girl for life.

*Inhumane physical abuse leading to disabilities suffered by 5 month old baby
John who nearly escaped death*

John is now 6 years old living with his grandparents happily who is legally adopted by them. He is not a normal child as his father left with him with scars and disabilities for all his life. He is suffering from seizures, delayed development, nerve damage in his ankle, right sided weakness, sensory disorders, autism and half blindness. He went through a painful journey of undergoing 22 surgeries which exceeded around $2 million for repairing brain damage, infection as well as removing brain matter.

When John was born in 2009, he was a healthy and normal baby. Both his parents Penny and Jack steward were working in shift's so one or the other were always present to take care of the child needs. Penny's parent's happiness was boundless by the birth of their grandson. When they visited their 1 month old grandson they suspected that penny and jack were neglecting their child's needs. Forget about needs, they didn't even have time to love their only son and seemed detached. They didn't even have time to change their son's diaper and feed him regularly. Baby john was kept in a separate room and lights were always switched off. Sometimes baby john used to cry all night in hunger and fear and none of his parents would wake up to tend to his needs.

On June 2011, the grandparents were informed by penny that baby john is having vomiting, seizures and injuries due to falling down from stairs. They immediately reached hospital and were shocked to learn that their son-in-law is being arrested for physical abuse and trauma on baby john. Doctors told John's grandparents that it was a miracle that the baby is alive.

John was in an unconscious state. His grandparents were heartbroken to see their grandchild wrapped with bandages around his head, face and several wires stuck into his injured body. His health seemed deteriorated and was only 15 pounds! Their frustration and anger increased further when they learnt from doctors report that baby john was slammed down repeatedly which caused laceration on his mouth & chin, bruises, heavy injury on back of his head, brain injury, broken ribs and retina hemorrhage in right eye.

Investigations revealed that 1.6 year old baby john's father had beaten him up regularly due to frustration. When the baby was not fed properly he was crying due to hunger and in a rage his father pushed john backward which caused injury in his head, then he also slammed his son against the wall in anger. At that time penny was at work and when Jack saw blood oozing out of his son's head he immediately called his wife.

In the hospital doctors were not sure whether john will survive this trauma. This was said to be the worst baby abuse cases treated by them. But his grandparents were confident that he will survive and they adopted him immediately and swore to look after him and love him for their lifetime. Jack steward is now serving 18 years in correctional facility and his wife was also sentenced to 4 years of prison for child neglect.

Even though John survived and is living in a pleasant environment he will never be free from traumatic disabilities caused by his father. His misshaped skull will be a remind everyone of his unfathomable pain and suffering.

Dana, 3 year old toddler locked and raped by father repeatedly for 4 years

Dana was only 4 years old when she could hear from her closet two of her brothers and sister laughing and playing outside who were not being abused. She was tortured, raped, beaten and starved by her father for 5 years. The little girl was always in constant fear and confusion. Astonishing and unacceptable fact is that her mother was also enjoying this ordeal. The baby girl was used by her own father to satisfy his cruel desires.

One day Dana's father got so angry by her sobbing that he pushed her inside the cupboard and closed it. The little girl thought that she was being punished and cuddled under the dresses hanging in the closet in total darkness. Even her mother was not bothered about her. She was just happy taking care of her other three children. Dana was not given any food or water for hours. When she was becoming unconscious her father pulled her out of the closet and raped her. She was just 3 years old! She was filled with agony and in a confused state. This continued for years. That cupboard became her new home where she had to sleep and also use it for bathroom. The bed sheet under her was always wet with her urine and had to sleep under a wet blanket. She was also beaten brutally and burnt with cigarettes.

Dana became 6 year old and was suffering from malnutrition. She was tied down in the closet so that she wouldn't fight back while she was being sexually abused by her father. It was as if she was left alive for suffering. On her brother's birthday Dana was given cake by her father but he was not letting her eat it. He told her to chew the cake and spit it on the floor. Her mother would push her in bathtub when her siblings went to school. She also tried to drown dana in hot water whenever she was stressed out.

One day a neighbor came by to borrow something and luckily Dana's mother forgot to lock the main door before going for shower. The neighbor sneaked in just for fun but when she saw what was inside the cupboard she freaked out at the state of 8 year old Dana. Immediately officials were called by the neighbor and dana was rescued from her misery putting an end to it.

Everyone thought that it was a miracle that she was alive after going through such horrific abuse by her father. Dana was saved by the doctors even though her organs were failing. She was put on surgery and with extensive care her body organs started responding again. After she came out of danger Andrew and Mary Hopkins adopted her. Both her abusive parents were arrested and convicted of physical abuse sentencing them to life time imprisonment.

It took time for Dana to trust anyone. Even though her adoptive parents were very caring & loving towards her she was traumatized mentally and physically. She was in constant state of fear and would often hide her food thinking that she might live in the closet again. Mary was speechless when she saw that dana didn't even know how to play with anyone or even with toys. Dana's child specialist told them that she has the mental state of a 3 year old. Sometimes she couldn't sleep in normal bed and went inside any closet and slept there peacefully for hours. She was scared of taking bath as she scream, "please don't drown me, I didn't do anything".

With proper care and love given by her guardians dana recovered from her state of traumatic condition and went back to school to complete her studies. Brave Dana, now 21 year old has recovered from trauma is now helping other victims of physical abuse.

Baby girl Riah brutally assaulted and mistreated by her father causing long term trauma

5 month old baby Riah had been suffering physical abuse for months before she was admitted by her parents with severe injuries and brain damage. Before she was hospitalized the baby girl had suffered seizures and injured ribs caused by her abusive father. She is now 15 months old suffering from brain damage and remains in a vegetative state.

Based on medical reports doctors stated that riah had been suffering from traumatic injury one week earlier and her health started deteriorating. As she was not given treatment on time her body was not responding well to stimulus. She also had brain injuries, broken bones and fractured ribs which were healing gradually.

Riah's father Anderson Gibbs was often the sole caretaker when her mother patty was attending college. They were living with patty's mother who always confronted them about the baby girl's injuries and her lethargic state. But anderson was not bothered and he was always abusive towards everyone. Neighbors and relatives also found photographs of baby riah with injuries in left eye, lips, chin and head. In one photo her head was bulged as if she was slammed hard. But nobody was sure about the abusive activities suffered by riah due to her father's temperament.

After baby riah was admitted in Child care hospital her father was arrested and was being held on $300,000 bail. He refused ever abusing his child. So investigators seized his phone to gather proof. They found several photos and videos of the baby having injuries and seizures. His wife also confessed about his abusive nature towards her little girl. She claimed that anderson used to strangle riah from back of her neck pushing her face down on the ground in the bathroom. Her daughter was also grabbed by her father brutally which caused several bone injuries.

With proper evidence Anderson was charged with first degree assault and mistreatment of his baby girl. Riah, now 15 months old was taken into custody and placed with her grandmother.

Emotional and physical abuse suffered by 5 year old Jerry for lack of concentration in studies

Jerry, 5 year old was not so good in his studies. His teachers thought that he was not putting many efforts as he was struggling to get average marks. But he was not like that in the beginning. His mother Kate Watson claimed that her boy was very sharp and was very much interested in his studies. Due to his father Barry's bullish nature jerry was so scared to study anything.

Instead of teaching his child properly jerry's father would regularly beat him up brutally and even hang him if his answers went wrong. Kate was humiliated too if she came to rescue her child. So she took her cellphone and got a video of abusing his child without barry's knowledge. The horrifying video showed that the 5 year old boy was studying in his room; each time he gives any wrong answer to his father he kicks him and slams his head on the desk. This went on for some time and the boy was so scared that his hands started shaking and his dad became impulsive with anger and loses his temper. By the time the boy could even shout for help barry strangles his neck from behind and hurled him to the wall before the boy started with seizure.

The little boy was rushed to the hospital and his mother immediately went to the officials to give them the video. Even the investigators said that they have never seen such disturbing act of violence against any child by their own parents. This video was also shared in facebook for more than 300,000 times to spread awareness against child abuse. Barry was arrested and found to be a drug addict. He was charged with illegal usage of drugs, emotional & physical abuse on child.

Jerry now 6 years old still remains unable to speak and unresponsive due to his traumatic past inflicted upon him by his own father.

Heartless Dad tried to sell his own 3 month old daughter

Abby gave birth to a beautiful daughter in the month of July 2013. After her birth she died due to complications in surgery. Jaden was left alone to raise his baby girl. She was named by her grandparents as 'Ashley'. Her father was not happy with his daughter and blamed his small child for the death of his wife. He was addicted to drugs and drinking problems.

Little Ashley was being taken care by her grandparents. They were not aware that Jaden needed psychiatric therapy to come out of his misery. He was not able to think clearly. Whenever he went to look after his 2 month old daughter he had an urge to drown or kill her. Grandparents noticed several reddish marks around their grandchild's hands and legs. Once they caught her father trying to place her on the stove, but he lied that he was just trying to make his daughter warm.

Jade started hanging out with wrong people who were into child trafficking. They proposed to him that he could sell his daughter for $5000.

Jade didn't even think before doing this kind of unacceptable sin against his own 3 month old daughter.

Next day Jade told his parents that he is taking Ashley for vaccination and took her in a hurry. His parents became suspicious of his behavior. They followed him and found him exchanging their grandchild with a stranger in the corner of a restaurant. Soon Ashley's grandfather rushed to her rescue and snatched her from the stranger's hands. Nearby officials became alert and arrested Jade along with the child trafficker.

Ashley was admitted in the hospital and was found to be having several internal injuries around her body. Doctors said that this kind of injuries occurs only when someone tries to crush a baby intentionally. Jade was sent to psychiatric treatment for 2 years and imprisoned for 5 years for child neglect. Police blamed his parents that they should have given jade proper treatment when he started abusing his newborn child.

9 month old baby girl tormented physically and carried injuries which left her paraplegic

When Abigail was just 9 months old, she was admitted to Child care hospital of Florida. She carried with her battered child syndrome, multiple fractures in backbone, retinal hemorrhages and severe injury in spinal cord. Her medical report showed that she would impair in her legs for lifetime.

Both her parents were arrested for child neglect, severe physical harm and suspected murder. One year old Abigail was placed under the care of social worker Caitlin and she adopted her legally at the age of 2. Even though baby girl Abigail was under good care, she was suffering from disability and development delays.

Before she was admitted in hospital Abigail was suffering severe abusive and life threatening behavior from her father Hank Wiley. She was being spanked regularly whenever she cried for milk or change of diaper. One of the neighbors quoted that she could hear Hank yelling at the baby and abigail's cried for a long time. Some even claimed that her father would shake her to make her sleep.

Abigail's mother was working as a beautician in a parlor. It was as if her mother was also not interested in the child's development or her basic needs. All she wanted was more money. To manage her own needs she was working overtime and was always hanging out with friends. Hank on the other hand who had suffered traumatic childhood due to his parents strict behavior he was mentally affected. Even though he loved his daughter, he was not able to control his anger and showed his frustration on his daughter by beating and spanking his baby girl while tending to her needs.

One day Hank crossed his limit and beat his daughter severely on her back and she started screaming in agony. She was rushed to the hospital and doctors said that her spinal cord is injured severely disabling both the legs. Her father was sentenced to 10 years of imprisonment and mother to 4 years of imprisonment for not giving proper care & child neglect. Abigail was going through post traumatic disorder, sleeping problems and hyperactivity.

Now 9 year old Abigail is enjoying a great and healthy life with her guardian. Even though she is not able to walk her future looks promising because of her confidence and courage. She even wrote a sweet note to her guardian stating that she is an angel in her life who was sent by god to make her life beautiful.

Daughter sexually abused by father repeatedly for 2 years

Lacey was sexually abused by her father between the age of 6 to 8 years. When she disclosed to her mother Claire about her father's abuse she was ignored & rejected. Her father Houston told Claire that their daughter was imagining things from seeing adult movies. And Claire stopped talking to her daughter properly and always abused her emotionally.

6 year old Lacey was victim of emotional and sexual abuse by her own parents. She started to have eating and sleeping disorders. Whenever her mother went for shopping she was left alone with her father. He would force her and rape her by slapping on her face. Houston even made a video of raping her and shared with his cunning friends.

She was confused and always felt agony while Houston touched her private parts while raping her. She felt as if she wanted to die. Nobody believed her which led her to severe depression. Lacey was hospitalized for attempting suicide by hanging herself with a rope in her bedroom.

When her story came into light, Houston ran away & was hiding in one of his friend's house. Medical reports showed that Lacey was also going through internal bleeding due to severe penetration and sexual abuse by her father. Claire felt heartbroken and was not able to look at her daughters eyes for not believing her. Lacey's father was found by the police and sentenced to life time imprisonment.

It took lacey 4 years to recover from her internal injuries and mental trauma. Before her recovery she was going through poor self-esteem, depression, anxiety and suicidal behavior. She even thought herself to be unworthy of living. She was closed from inside and felt nothing.

Lacey, now 11 years old is still recovering from her horrific and painful events caused by her father.

6 year old son physically abused by father every day for 2 years leaving young Preston traumatized forever.

Gary Darwood, 30 was a software engineer. His wife Wanda was a housewife and an affectionate woman. Their son preston was a joyous and cheerful boy in the neighborhood, until his father addicted himself with gambling and drinking problems. Their happy life was turned over within a night into disaster.

Preston was good in studies too. He liked painting, cooking and playing. All of a sudden his father came under wrong influence and everything changed forever. His drunken father would come home and abused his son & wife every day. They became poor due to his gambling addiction. He lashed out all his frustration on his son and wife by beating them regularly. This went on for days.

Once Wanda went to visit her parents to take care of her father who was a heart patient. As usual Gary came home fully drunk and verbally abused his son that he has to spend so much of money on his studies which is a waste. He grabbed preston twisting his ears and forcing him to kneel down to spank him. His father went on beating him and didn't stop even when his child was screaming & begging to stop in agony. Preston was slapped several times and kicked in his stomach without any pity. As he was mistreating his boy, Preston's right hand & fingers was fractured. Hearing his painful screams the neighbors immediately called the police and they arrested his father Gary Darwood.

Preston was immediately admitted in Emergency hospital and given treatment. Doctors found abdominal injuries, severe injuries all over the body, broken fingers & fractures in the right hand. Even though he was given proper treatment and cured most of his injuries. He showed signs of breathing problems and movement in his right hand fingers was difficult for him.

Preston became victim to his abusive father due to his addiction to alcohol. His father was sentenced to 10 years of imprisonment and he was ashamed of his actions towards his son.

After this horrific incident preston was not normal. He was going through depression, anxiety, poor school performance and inferiority complex. The physical abuse and violence caused by his father made him fragile. With the help of his mother and family members preston is trying to recover from his trauma. But he is still not able to write or draw properly due to his injury in his right hand & fingers.

Our children suffer unacceptable consequences for their father's unjust actions. Even the neighbors or his mother could have prevented abuse by resisting any kind of violence towards their children.

www.ingramcontent.com/pod-product-compliance
Lightning Source LLC
Chambersburg PA
CBHW072253200526
45168CB00015B/1737